A Boy is a Boy
A Girl is a Girl

A Rhyming Story of Reality

Written By Stephanie Fields
Illustrated By AJ Hagen

Written by Stephanie Fields

Illustrated by AJ Hagen

Edited & Designed by Nicole Mullaney

Published by SF Publications

sfpublications369@gmail.com

Hardback ISBN: 979-8-9916717-5-0

Paperback ISBN: 979-8-9916717-6-7

Ebook ISBN: 979-8-9916717-7-4

Dedication

To the parents, grandparents, and guardians who refuse to let common sense become a fairy tale. Thank you for being the anchors in the storm. This book is as much for you as it is for them.

The Beginning

From the very first day,
when the world said, *Hello*,

There were things about you
that were certain to grow.

With a wiggle of toes and
a tiny button nose,

You were a boy or a girl,
from your head to your toes.

The Design

The birds are all birds,
and the bears are all bears,

The lions have manes. The sheep
have their wool and hairs.

Nature has patterns,
a wonderful plan,

To grow a little lady or
a sturdy young man.

An apple seed grows to
an apple tree tall,

It doesn't turn into a plum
tree at all!

A kitten stays a kitten,
a pup stays a pup,

And you stay YOU as
you're growing right up.

Feelings and Fun

You might like to climb,
or you might like to sew,

You might like the fast way,
or prefer to go slow.

A girl can love engines,
a boy can love art,

But that doesn't change how
you were made from the start.

If a boy plays with dolls,
he is still just a boy,

He's simply a person
enjoying a toy.

If a girl wears a tie, suit,
or a hat,

She's still just a girl,
and that is just that!

The Inside Secret

Deep down in your body,
in places unseen,

Is a code that is perfect,
honest, and clean.

It's called DNA, and it's
there to stay,

It tells the whole world
who you are every day.

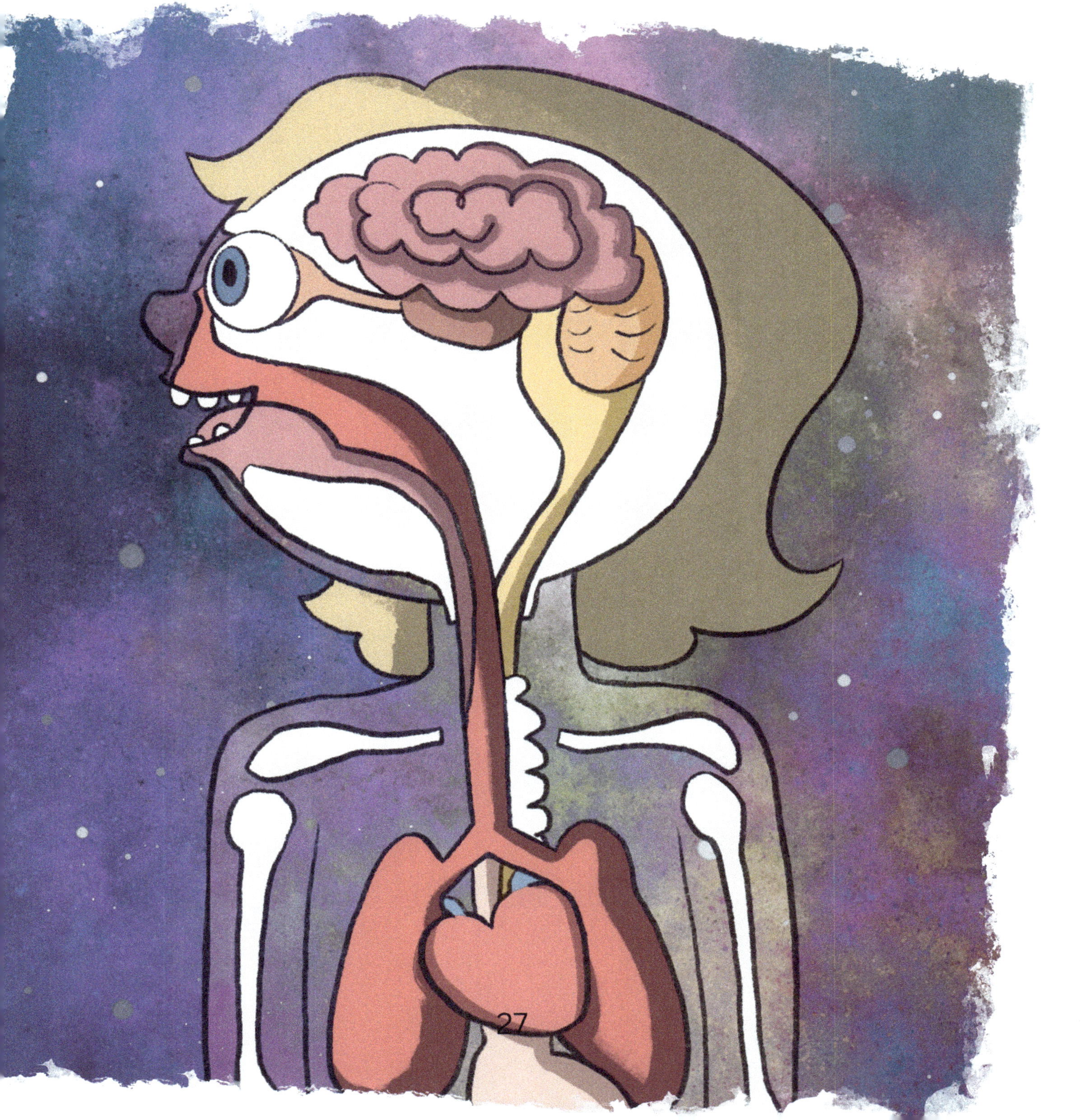

It's like a secret signature,
written so small,

It's the truest thing about
you, the best thing of all.

It says you're a daughter,
or you're a son,

A beautiful truth from the
day you begun.

The Promise

So look in the mirror
and give a big grin,

Be proud of the body
that you're traveling in.

You don't have to choose
or guess who you are,

Like picking a toy or a wish
on a star.

You were made with a purpose, reason, and light.

Being who you are made as is perfectly right.

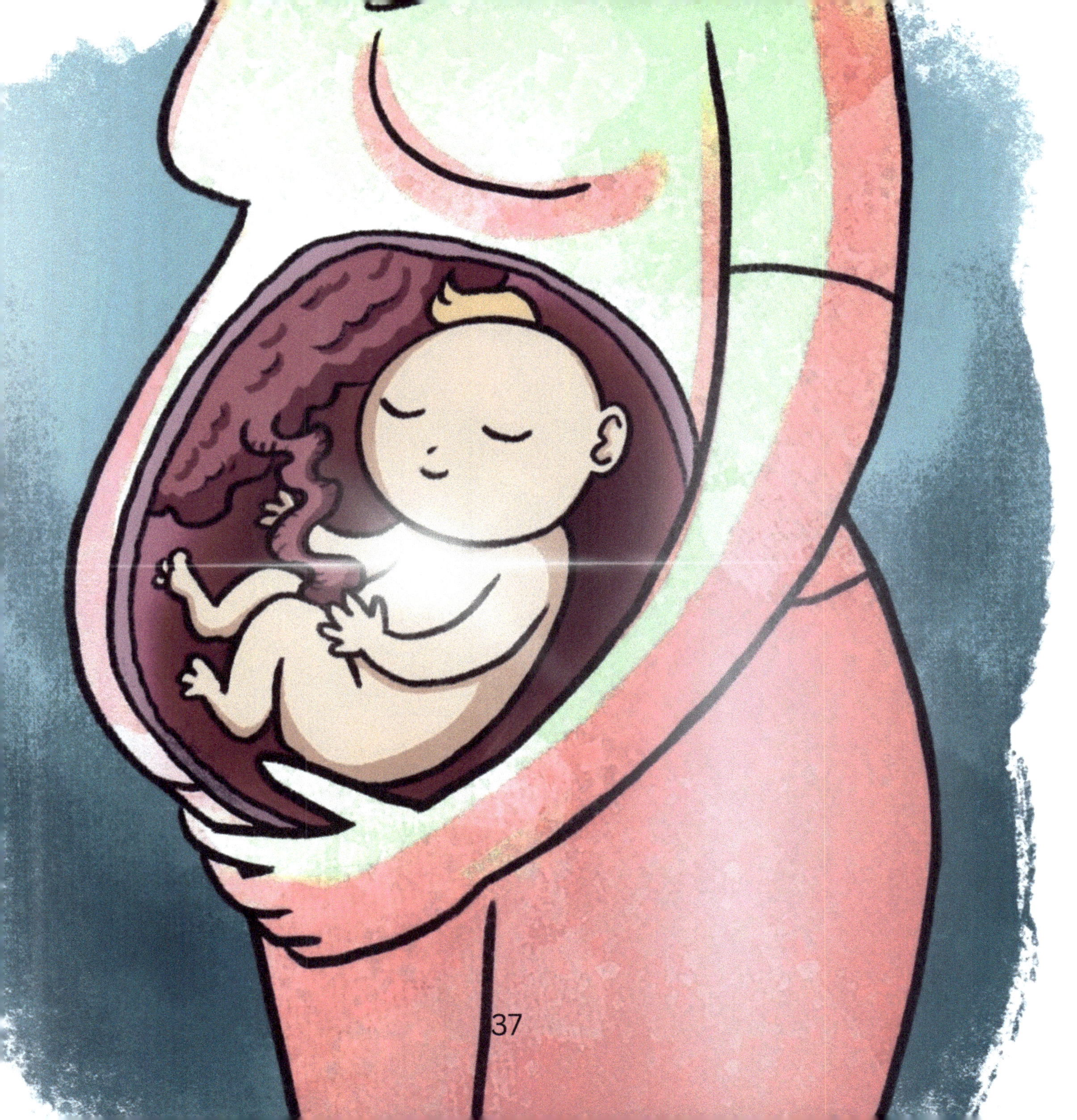

So stand very tall,
with your head in the air,

The truth is a treasure
that's always right there!

The Discovery Page

Interactive Nature Walk

Parents: Read these aloud and let your child point to the answers!

Nature always tells the truth! Can you find the match?

1. The Little Nest: A mother bird sits on her eggs. When they hatch, will they be baby robins or baby goldfish?
   - (Answer: Baby Robins! Animals stay true to their kind.)
2. The Garden: We planted a sunflower seed in the dirt. Will it grow into a giant sunflower or a prickly cactus?
   - (Answer: A Sunflower! Seeds grow into what they were made to be.)
3. The Mirror: Look at your reflection. If you put on a tiger mask, do you actually turn into a real tiger?
   - (Answer: No! Costumes are for fun, but the person underneath is still YOU.)

The Big Talk: Q&A for Kids and Parents

Simple answers for big questions.

Q: "Can I still be a girl if I like to play in the mud and play with trucks?"

A: Of course! Being a girl isn't about the toys you like or the messes you make. You can be a "tomboy" and still be 100% a girl just as you were born as. Your body tells the truth, not your toy box!

Q: "If a boy paints his nails, does that make him a girl?"

A: Nope! Painting your nails is just like wearing a colorful shirt. It’s a choice for your style, but it doesn't change your DNA. He is still a boy, just with colorful fingers!

Q: "Why do some books say I can pick if I'm a boy or a girl?"

A: Sometimes people get confused and think our feelings are the boss of our bodies. But our bodies are a gift that tells us the truth. Just like we can’t choose to have wings and fly, we don't choose if we are a boy or a girl. We are born that way, and that is a great thing!

My Truth Page

I am Wonderfully Made!

the box below, draw a picture of yourself doing something you love. Are you running? Drawing? Helping?

Cut this out or sign it right here in the book!

The Certificate of Wonderful You

OFFICIAL CERTIFICATE OF REALITY

This belongs to:

____

I am a [BOY / GIRL]

I was made with a special design from the very start.

I am proud of my body, DNA, and heart.

I don't have to change to be special or "new"

I am happy just being the biological ME!

Signed: ______________________________ (Child)

Witnessed by: __________________________ (Parent)

Date: ______________________________

A Note from the Author

We did it!

To my incredible audience, community, and every single one of you who hit that "Pre-order" button: Thank you!

When I made the choice to stand up for common sense and protect our children, I knew I couldn't do it alone. I bypassed the traditional gatekeepers and put my trust entirely in you, and you showed up louder and stronger than I could have ever imagined. Your overwhelming support didn't just fund this book, you sent a clear message that our children's minds are worth fighting for. You are the backbone of this movement, and I am so deeply honored to walk this path with you.

This book belongs to all of us. Please be sure to visit the Hall of Acknowledgments page to see the names of the incredible patriots who stepped up and donated to make this project possible!

To My Family

To my wonderful family: thank you for being my rock throughout this entire journey. Thank you for your endless patience, letting me read this book out loud to you, and always giving me your honest feedback and endless support. Your love kept me grounded, your critiques made the story stronger, and knowing you were in my corner gave me the courage to put this message out into the world.

To My Creative Team

AJ Hagen, my brilliant illustrator: Thank you for your tireless hard work and breathing life into these pages. You took the words of this story and transformed them into beautiful, vibrant art that children will cherish for generations. Your talent is a gift, and this book simply wouldn't be what it is without your vision.

Nicole Mullaney, my editor and formatter: Thank you for your sharp eye, dedication, and incredible skill in polishing this manuscript and structuring it so beautifully. You helped me center this message perfectly, ensuring that every rhyme, activity page, and parent guide flows seamlessly.

We stood our ground, kept the faith, and together, we are putting truth back into the hands of the next generation.

With all my love and gratitude,

Stephanie Fields

Don’t forget to leave a review! It helps more than you know!

About the Author

Stephanie is a dedicated advocate for children who brings over a decade of professional experience as a former teacher and social worker to her writing. While she no longer serves in those roles, her years spent in classrooms and supporting families deeply informed her understanding of how children process big ideas.

Today, Stephanie's mission has moved from the field to the page. She realized that the most effective way to help children navigate modern questions was through the power of engaging rhymes and high quality, colorful illustrations. By combining her professional background with her mom-heart, she creates resources that are as fun to look at as they are foundational to hear.

As a parent of children in the public school system, Stephanie saw firsthand the confusion modern cultural shifts were creating. When her own kids began asking difficult questions about identity, she saw a need for a resource that offers simple, loving and biological truths.

This book is her answer, a way to secure the shield for families everywhere. Stephanie lives in Texas with her husband and their children, where she enjoys photography, creating and reading.

About the Illustrator

Bringing the world of this book to life with her professional illustrations, is AJ Hagen, with over a decade of experience in the industry. AJ holds a BFA in illustration, and her technical expertise is evident in every vibrant, high quality character she creates, from our majestic lion to the gentle lamb.

Her career has been dedicated to visual storytelling that captures the imagination of little eyes while maintaining a sophisticated, professional polish. AJ specializes in creating warm, engaging environments that make children want to revisit the pages again and again.

AJ lives in Michigan with her husband and their children. When she isn't at her drawing desk, she’s busy raising her own brood of explorers, bringing a mother's perspective to every brushstroke.

Hall of Acknowledgements

When I started this project, I expected challenges, but I never anticipated a road this wild!

We hit major speedbumps very quickly out of the gate. Kickstarter suspended the project and flagged it as "unsafe" after a wave of reports from the opposition. Then, when I went to TikTok to share what happened, the algorithm suppressed my videos to only 9 views. It really felt like every step forward came with two steps back.

But I refused to give up.

Despite the pushback and setbacks, my team and I carried on because this work is too important to stop. Thank you for standing by me through the noise and believing in this journey. We made it through the storm, and I'm so excited to finally share this with you!

Thank you to the following supporters who showed up and showed out for "A Boy is a Boy, A Girl is a Girl". Please, give them a round of applause!

Jody Mooneyham backed this project by $1,250!

$100 Backers

Madeline Writes
JeremiahOneFive
Susan Greely
Jerika Evors

$50 Backers

Amelia Grace Jones
Mistybuki
Cathi Baugh
Jipsi Harris
Ali Carson

Thank you to those who showed support by sharing the project to their following

DE Carlson
SassySouthernPatriot

Let's Connect

You can find Stephanie Fields
on the social media accounts below.

Tiktok - @stephaniefieldsauthor

Facebook - @Stephanie Fields
(you'll see my books)

Instagram - @stephaniefieldsauthor

X - @Stephanie Fields Author

You can find A.J Hagen
on the social media accounts below.

Tiktok @nerevarine88

Instagram @artofajh

www.ingramcontent.com/pod-product-compliance
Lightning Source LLC
LaVergne TN
LVHW070443061126
851858LV00011B/156
9798991671750